Congestion Pricing:
A PRIMER ON INSTITUTIONAL ISSUES

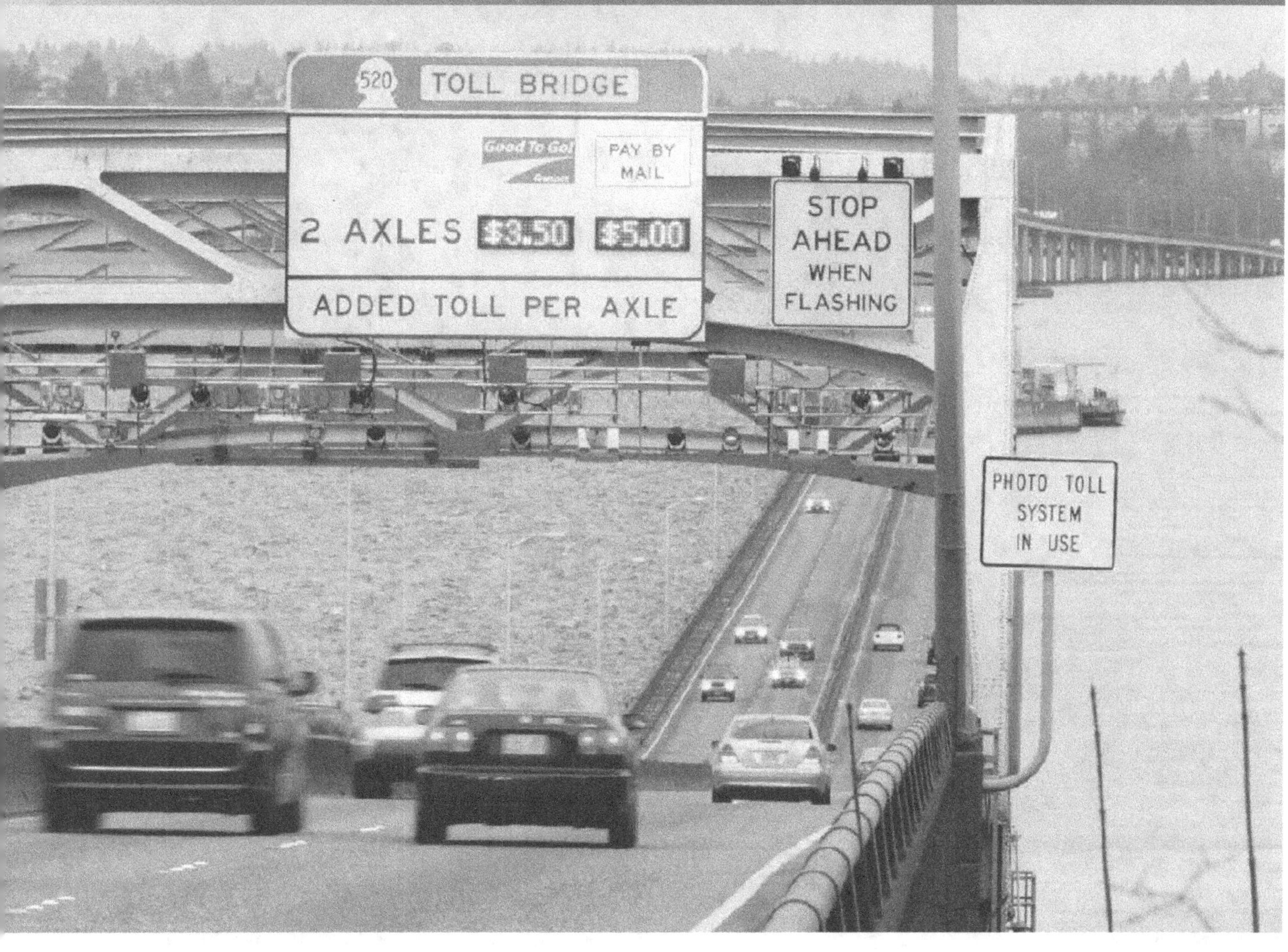

Notice

This document is disseminated under the sponsorship of the U.S. Department of Transportation in the interest of information exchange. The U.S. Government assumes no liability for the use of the information contained in this document. This report does not constitute a standard, specification, or regulation. The U.S. Government does not endorse products of manufacturers. Trademarks or manufacturers' names appear in this report only because they are considered essential to the objective of the document.

Quality Assurance Statement

The Federal Highway Administration (FHWA) provides high quality information to serve Government, industry, and the public in a manner that promotes public understanding. Standards and policies are used to ensure and maximize the quality, objectivity, utility, and integrity of its information. FHWA periodically reviews quality issues and adjusts its programs and processes to ensure continuous quality improvement.

Technical Report Documentation Page

1. Report No. FHWA-HOP-13-034	2. Government Accession No.	3. Recipient's Catalog No.
4. Title and Subtitle Congestion Pricing: A Primer on Institutional Issues	5. Report Date April 2013	
	6. Performing Organization Code	
7. Author(s) Donald Samdahl (Fehr & Peers), Myron Swisher (SAIC), Jennifer Symoun (SAIC), Will Lisska (Fehr & Peers)	8. Performing Organization Report No.	
9. Performing Organization Name and Address Science Applications International Corporation (SAIC) 11251 Roger Bacon Drive, 3rd Floor Reston, VA 20190 Fehr & Peers 1001 4th Avenue, Suite 4120 Seattle, WA 98154	10. Work Unit No. (TRAIS)	
	11. Contract or Grant No. DTFH61-06-D-00005	
12. Sponsoring Agency Name and Address United States Department of Transportation Federal Highway Administration 1200 New Jersey Ave., SE Washington, DC 20590	13. Type of Report and Period Covered	
	14. Sponsoring Agency Code HOTM	
15. Supplementary Notes COTM: Ms. Angela Jacobs, Federal Highway Administration		
16. Abstract Institutional issues provide challenges to implementing congestion pricing strategies. The primer explores the types of institutional issues that are commonly encountered with priced lanes, zone-based pricing, and parking pricing programs. These issues include the up-front challenges of establishing leadership, meeting legislative requirements, and setting an organizational structure. Once the challenges are overcome, there are numerous institutional issues related to the planning process, public involvement and implementation procedures. The primer examines these topics with the insights gained from case study applications around the United States and Europe.		
17. Key Words Road pricing, congestion pricing, priced managed lanes, parking pricing	18. Distribution Statement No restrictions.	
19. Security Clasif. (of this report) Unclassified	20. Security Clasif. (of this page) Unclassified	21. No. of Pages 40	21. Price N/A

Form DOT F 1700.7 (8-72)   Reproduction of completed page authorized

Cover image source: Washington State Department of Transportation (DOT)

Contents

The Primer Series and the Purpose of This Volume	1
Introduction	3
Congestion Pricing Strategies	3
Types of Institutional Issues	4
Organization of the Primer	4
Leadership	5
Legislative	8
Organization	12
Planning Process	16
Public Involvement	20
Managing Costs and Revenues	23
Implementation	26
Lessons Learned	29
References and Resources	30
References	30
Resources	30

The Primer Series and the Purpose of This Volume

States and local jurisdictions are increasingly discussing congestion pricing as a strategy for improving transportation system performance. In fact, many transportation experts believe that congestion pricing offers promising opportunities to cost-effectively reduce traffic congestion, improve the reliability of highway system performance, and improve the quality of life for residents, many of whom are experiencing intolerable traffic congestion in regions across the country.

Because congestion pricing is still a relatively new concept in the United States, the Federal Highway Administration (FHWA) is embarking on an outreach effort to introduce the various aspects of congestion pricing to decision-makers and transportation professionals. One element of FHWA's congestion pricing outreach program is this Congestion Pricing Primer series. The aim of the primer series is not to promote congestion pricing or to provide an exhaustive discussion of the various technical and institutional issues one might encounter when implementing a particular project; rather the intent is to provide an overview of the key elements of congestion pricing, to illustrate the multidisciplinary aspects and skill sets required to analyze and implement congestion pricing, and to provide an entry point for practitioners and others interested in engaging in the congestion-pricing dialogue.

The concept of tolling and congestion pricing is based on charging for access and use of our roadway network. It places responsibility for travel choices squarely in the hands of the individual traveler, where it can best be decided and managed. The car is often the most convenient means of transportation; however, with a little encouragement, people may find it attractive to change their travel habits, whether through consolidation of trips, car sharing, by using public transportation, or by simply traveling at less-congested times. The use of proven and practical demand-management pricing that we freely use and apply to every other utility is needed for transportation.

The application of tolling and road pricing to solve local transportation and sustainability problems provides the opportunity to solve transportation problems without federal or state funding. It could mean that further gas tax, sales tax, or motor vehicle registration fee

---

About This Primer Series

The Congestion Pricing Primer Series is part of FHWA's outreach efforts to introduce the various aspects of congestion pricing to decision-makers and transportation professionals in the United States. The primers are intended to lay out the underlying rationale for congestion pricing and some of the technical issues associated with its implementation in a manner that is accessible to non-specialists in the field. Titles in this series include:

- Congestion Pricing Overview.
- Economics: Pricing, Demand, and Economic Efficiency.
- Non-Toll Pricing.
- Technologies That Enable Congestion Pricing.
- Technologies That Complement Congestion Pricing.
- Transit and Congestion Pricing.
- Income-Based Equity Impacts of Congestion Pricing.
- Congestion Pricing Institutional Issues.

The primers are available on the FHWA Congestion Pricing web site: http://www.ops.fhwa.dot.gov/tolling_pricing/index.htm.

increases are not necessary now, or in the future. The idea of congestion pricing is a conceptual first step, not a complete plan of action. It has to be coordinated with other policy measures and environmental measures for sustainability.

Institutional issues provide challenges to implementing congestion pricing strategies. This primer explores the types of institutional issues that are commonly encountered with priced roadway lanes and parking pricing programs. These include the up-front challenges of establishing leadership, meeting legislative requirements, and setting an organizational structure. Once these challenges are overcome, there are numerous institutional issues related to the planning process, public involvement and implementation procedures. These topics are examined with the insights gained from case study applications around the United States and Europe.

Introduction

CONGESTION PRICING STRATEGIES

There are five main types of congestion pricing strategies aimed at shifting travel demand away from peak period travel and/or to alternative travel modes:

1. *Priced lanes*: Variable or dynamically priced tolls on separated lanes within a highway, such as express-toll lanes or high-occupancy toll (HOT) lanes.

2. *Tolls on entire roadways*: Tolls on roads or bridges, including tolls on existing toll-free facilities during rush hours

3. *Zone-based (area or cordon) charges*: Either variable or fixed charges to drive within or into a congested area within a city.

4. *Area-wide charges*: Per-mile charges on all roads within an area that may vary by level of congestion.

5. *Parking Pricing*: Use of parking pricing to modify travel behavior in terms of schedule, mode, and/or location. Can also include a range of parking cash-out policies in which cash is offered to employees in lieu of subsidized parking

Most congestion pricing strategies have some common objectives, such as managing peak period traffic demand and encouraging use of alternative modes. However, there are some notable differences in the strategies that have bearing on the institutional issues. Some of these differences are highlighted below.

- **Priced lanes** – Most applications of priced lanes are in the form of HOT lanes on freeways.[1] These facilities are typically under the control of a State department of transportation (DOT) or a toll authority, resulting in relatively clear lines of authority in terms of planning, design, and operation. In addition to the toll-paying drivers, users of the HOT lanes typically include transit agencies, carpool/vanpools, and emergency vehicles. These users may have different objectives than those of the implementing agency. The advent of Public Private Partnerships (P3) has also introduced the private sector into the mix. To date, HOT lanes have been implemented in selected freeway corridors. Expansion into HOT lane systems begins to expand the institutional complexities.

- **Zone-based pricing** – The only examples of zone-based pricing are outside of North America. Typical applications are cordon-pricing around a city center. While one agency may have overall responsibility for the project, there are typically many other agencies involved, since the project affects a large number of roadway facilities and services. The private sector (e.g. businesses, residences) become active participants especially at the boundaries of the zone.

- **Parking pricing** – The parking pricing strategies examined in this primer focus on government agencies adjusting on- and off- street parking prices to influence the timing and location of parking within a city. As such, the institutional issues tend to be focused within the city, including its interactions with private-sector parking operators and affected businesses.

---

[1] The SR 520 bridge replacement project in Seattle is an example of a variable tolls being charged on an entire freeway facility.

TYPES OF INSTITUTIONAL ISSUES

Implementing congestion pricing projects requires a combination of good planning, proven technical capabilities, and reliable day-to-day operations. Equally important is the ability to work through the myriad of institutional issues that can arise at any point in the process. Congestion pricing projects are different from many other transportation projects: they represent a new way of managing travel demand; they require daily, hourly, or even constant monitoring; and they deal explicitly with money. All of these factors produce an inordinate amount of attention by decision-makers, the media, and the public.

This primer will explore the range of institutional issues that have arisen on congestion pricing projects throughout the United States and Europe. These issues were examined through a variety of sources: interviews with practitioners, an FHWA-sponsored webinar, and a peer exchange.[2] Through this research, the institutional issues primarily fit into the following types:

- Leadership – Project champions, roles and responsibilities
- Legislative – Enabling legislation, political structure
- Organization – Internal and external structures, interagency agreements
- Planning Process – Setting objectives, agency coordination, setting policies, developing the plan
- Public Involvement – Outreach strategies, gaining public support, marketing
- Managing Costs and Revenues – Cost sharing, allocation of revenues
- Implementation – Construction and roll-out, day-to-day operations

While many of these issues are broadly applicable to the wide range of congestion pricing strategies, some are more unique to a particular strategy. This primer focuses on issues as they relate to three strategies: variably priced lanes, zone-based charges, and parking pricing (except parking cash-out) strategies. These three are not intended to be all inclusive, but represent a broad array of congestion project types and institutional issues. The first variably priced fully tolled facility (i.e. SR 520 bridge in Seattle) is also examined in the context of the other pricing strategies.

ORGANIZATION OF THE PRIMER

This primer is organized by the seven institutional issue types, with a section dedicated to each issue. Each section provides insights into how dealing with the respective issue directly within the project development process can lead to a more successful congestion pricing project. Each section also provides several strategy-specific examples illustrating how the issue has been handled in existing congestion pricing projects. The examples are color-coded by strategy as follows:

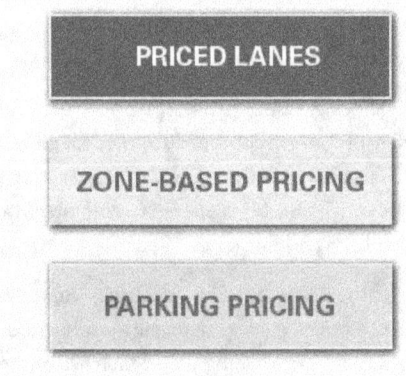

The primer concludes with a section on Lessons Learned, summarizing the key points identified for each institutional issue.

---

[2] Refer to the references for a list of interviews and other background material.

Leadership

A common theme among congestion pricing projects is the need for strong leadership to move the project from planning through design and into implementation. Because of the new and often controversial nature of congestion pricing in a community, the typical constituencies that support transportation projects may not be mobilized to move such projects ahead.

Leadership is manifested in two primary ways:

- Finding a Project Champion
- Developing strategic alliances

Key Points- Leadership

- The project champion can make the congestion pricing scheme effective
- Champions may include political and business leaders
- A project team or leader should seek to develop relationships with all of the key political and implementing organizations
- Agency alliances allow consistent messaging of project objectives and benefits

---

Leadership in the form of a Project Champion can help make the congestion pricing scheme effective. For roadway pricing projects, support is needed at both the political level and at the State DOT level. The project champion may often be a political leader; in some cases, leadership from major business groups can influence the project. A champion will typically be focused on the specific geographic application of congestion pricing (e.g. a HOT lane) and have a vested interest in its success. The project team can design the congestion pricing project to meet the project objectives, but the champion can keep the process moving ahead.

As discussed below, champions are also needed at the legislative level to ensure that sufficient enabling authority is provided to make the congestion pricing project a reality.

In addition to the project champion, forming strategic alliances can allow consistent messaging of project objectives and benefits. The project team or leader can develop relationships with all of the key political and implementing organizations so that there are few or no surprises. These organizations might include local and State DOTs, tolling authorities, transit agencies, and other operating authorities. Federal partners are also important on projects involving Federal-aid highways and/or Federal funding. Ideally, all agencies should seek to have the same priority with goals aligned among the organizations.

Champions Support Priced Roadway Lanes

Priced Roadway Lane projects have succeeded where there was clearly defined leadership and often a visible project champion. Here are some examples:

San Diego I-15 – The HOT lane project had a strong project champion, the City of Poway Mayor Jan Goldsmith. The mayor worked with other regional leaders to build support to convert the existing underutilized HOV lane on I-15 into a HOT lane. A key selling point was the ability to use some of the revenues to fund the operation of transit service in the corridor.

Atlanta – Leadership on the I-85 HOT lane project came from a strong relationship among three state and regional agencies who originally proposed on the UPA grant together (GDOT, the Georgia State Road and Tollway Authority, and the Georgia Regional Transportation Authority). This experience provided early bonding and trust building among these agencies, all of whom have been actively involved in Project Development and implementation.

Los Angeles – The Los Angeles I-10/110 HOT lanes project had strong support of the sponsor, Los Angeles Metro, but was not very popular with the California Legislature. This was due to concerns about social equity, traffic impacts, and public acceptance. State Senator Mark Ridley-Thomas, along with other public officials, emerged as a political champion when he authored the legislation that enabled tolling authority to make the HOT lanes a reality. The legislation includes a requirement for an assessment on the impact to low income commuters, a joint performance report from Metro and Caltrans at the conclusion of the demonstration period, and a formal public outreach and communications plan.

Seattle – In the Puget Sound Region of Washington State, there are several congestion pricing champions in both the public and private sectors. The Secretary of Transportation has been steadfast in support of tolling of the SR 520 bridge (i.e. a fully tolled facility), priced managed lanes on SR 167, and forthcoming express toll lanes on I-405. Also important is the leadership provided by major businesses such as Boeing and Microsoft, whose employees commute along these roadways. The private sector leadership has helped bolster legislative support to continue the pricing program.

Champions Needed for Zone-Based Projects

The influence of a project champion, or lack thereof, was evident in many of the zone-based congestion pricing projects, as described below.

The successful implementation of a congestion charge in **London** was due to Ken Livingstone, who was Mayor of London and head of the Greater London Authority (GLA). He provided political credibility and demanded technical credibility. The GLA had overall control (single tier government) and coordinated among other local authorities. Mr. Livingstone had full control to set the price and control the revenue. Three other zone-based programs were unsuccessful, in part due to leadership issues.

The proposed **New York City** program had a strong champion in Mayor Michael Bloomberg. It also was supported by both Governors who held office during this time. Strategically, Mr. Bloomberg acted in a similar manner to London's Livingstone, but unlike Livingstone he didn't have legal authority to implement a congestion pricing scheme. In this case, as project champion he could not overcome other political limitations outside of his control.

A proposed zone based program in **Edinburgh**, Scotland was hampered by a lack of strong local leadership. The decisions were dependent on Scottish government, which didn't have a local vested interest. As a result, when neighboring authorities were opposed, there wasn't a leader to help move the process along.

Finally, the project proposed in **Manchester**, UK had a good political structure to sponsor a zone-based congestion pricing scheme for the region. The region had one overarching agency (Transport for Greater Manchester- formerly the Greater Manchester Passenger Transport Authority) coordinating decisions for 2.5 million people combined with 10 elected authorities. However, the council chair was unable to convince the needed two-thirds of the member authorities to support the scheme amidst strong public opposition.

SFpark- An Example of Parking Pricing within a Defined Urban Center

The *SFpark* project in San Francisco is using parking pricing to manage congestion in the downtown. An advantage for San Francisco is that it is both a county and a city. While this can create some competing demands, decisionmaking is easier within this construct. The San Francisco Municipal Transportation Agency (SFMTA) plans and manages the transportation system for all non-auto modes. This includes control of on and off-street parking operations and enforcement. The SFMTA facilitated the roll-out of the *SFpark* program.

In San Francisco, *SFpark* leadership came through the city's political process and partnerships forged with nontraditional agencies and organizations, along with contacts between the city and private operators. The SFMTA Chief Financial Officer has been a vigorous supporter of the UPA program, under whose auspices the *SFpark* project was funded.

495 Express Lanes - Bringing in the Private Sector

The advent of Public-Private Partnership (P3) projects brings in the private sector. Public and private sectors often come into a project with very different perspectives and goals, and they negotiate differently. The agency must establish a strong relationship with the private concessionaire (and vice versa).

The Virginia I-495 Capital Beltway Express Lanes project was unique in that the HOT lane proposal from the private sector was unsolicited. As a result, there initially was no project champion from the public agency or political sectors. Once the design with HOV/HOT lanes and transit provisions was clarified, the Virginia Secretary of Transportation was in support. When a later Secretary created the Office of Transportation Public Private Partnerships (OTP3) in 2010, there was clear leadership on the overall P3 program.

Legislative

The legislative process forms the basis for proceeding with a congestion pricing project. Laws in many States (and cities) are silent on how to accommodate pricing within the transportation system. Agencies can plan all they want, but enabling legislation is usually required to move ahead with any congestion pricing project. This legislation defines the political structure within which the congestion pricing project can evolve.

"There WILL be a legislative issue. Everyone encounters it some way or the other"
(Minnesota Experience)

The operative word for gaining legislative support is "trust." Typically the sponsoring agency, for example a State department of transportation, needs to find support with key legislators, who must be able to trust that the project will be a success. This can be a "chicken-and-egg" situation. There are no guarantees of success, so the agency should choose its projects with an eye towards minimizing risk and maximizing the likelihood of success.

Good enabling legislation should provide clear authority among agencies and, if applicable, private parties. It should answer the question "who has control?" over decisions that need to be made throughout the project's evolution. Depending on the level of trust with the legislature, there may be more or less control given to the implementing agencies. Enabling authority doesn't necessarily mean that legislatures are willing to relinquish control.

The enabling authority should also help to define the purpose of applying pricing. For example, there is an inherent difference in toll structure if the objective is to maximize revenue rather than to manage traffic flow. Gaining some clear intent from the legislature can assist in designing a congestion pricing project that will meet political expectations.

There is substantial variation in political structures among regions in North America. The differences are more disparate when comparing North America to Europe, where most of the zone-based congestion pricing projects have been implemented.

Key Points

- Develop trust up front with legislature
- Need clear authority among agencies—determine who has control
- Have enabling legislation in place
- Seek overall authority without always needing to go back to legislature for decisions

Priced Managed Lane Legislative Experiences

With the growth in HOV to HOT lane conversions around the country, often the HOT lane planning process moved ahead of the legislative process. Different projects dealt with the enabling legislation a bit differently, with the end result being clear authorization to proceed with the HOT lane projects. A few examples:

Miami – As a recipient of a UPA grant, the Miami I-95 project took advantage of the fact that this was a discretionary grant with a short time frame required for completion. Therefore, some immediate legislative decisions were needed and the enabling process was expedited.

San Diego – The I-15 HOT lane conversion program originally had State representative support to gain approval in the legislature. San Diego Association of Governments (SANDAG) undertook extensive public outreach that helped lead to a political consensus to move ahead. Some legislative concern over the perceived "Lexus lane" syndrome was offset by committing funds to increased transit. During the legislative process, Caltrans was concerned with the fact that the project was originally a "demonstration" and pushed to set specific time frames for project implementation. This proved not to be a big issue and the project was approved.

Minnesota – Minnesota's initiation to priced managed lanes initially started with TRANSMART (a P3 process), which invited the private sector to propose and ultimately develop toll facilities. The State promised to provide substantial resources to match private sector funds to get the process started. Through this process, one project, State Highway 212, was advanced to a development agreement, but was ultimately vetoed under a legislative provision that allowed local communities through which a proposed toll facility passed to veto a project unconditionally. Despite the setback, the State commissioner of transportation at the time wanted to continue to pursue HOT lanes. The 1997 legislature finally authorized an HOV to HOT conversion of I-394 and, while exempting the project from the local veto authority in tolling legislation, did require a public hearing on the concept. This generated negative reactions from the public and politicians on perceived "equity issues" and the project was deferred. The Minnesota DOT's response was to form an external advisory task force guiding consideration and development of future projects. After a couple of years the task force supported the HOT lane concept and the legislature ultimately approved the I-394 project. By this time, a new Governor was also on board and became a champion for the project.

Los Angeles – The I-10/110 project was also a recipient of Congestion Reduction Demonstration program grant, which required the team to obtain enabling legislation in six months. The grant amount was so high that State elected officials had a hard time not passing legislation, however there was consternation with the short time-frame and the fact that not much planning had been done by that time. Some Congressional members also became involved, due to their concerns about tolling equity, diversion, HOV occupancy limits, and performance metrics. The project leaders did have to go back to legislature twice to revise the legislation, but both attempts were smoother because the team clearly defined the project design, demonstrated the expected benefits, and showed confidence in implementing the HOT lanes on time.

Washington State – The WSDOT initially received legislative authorization for tolling of the Tacoma Narrows Bridge. Subsequent authority was sought and received for the SR 167 HOV to HOT lane conversion and the SR 520 Bridge Replacement Project, which was part of the UPA grant program. SR 520 charges variable tolls on all lanes of the existing floating bridge. A strong public process has helped to build support for roadway pricing. A tolling implementation committee was established in 2010 to gauge and build public support and to provide guidance to the legislature. An initiative passed in 2010 requires the legislature to set toll rates. This additional requirement created challenges but was achieved. Currently all proposed tolling projects must be specifically approved by the legislature.

Virginia – The State already had P3 authority with the PPPTA Act of 1995. In 2002 the act was amended relative to design-build provisions. The original 1995 Act had good intentions to have the legislature make the up-front policy decisions, and then let the engineers and planners take over. In practice, the legislature was reluctant to relinquish control over the P3 projects (e.g. Hampton Roads Midtown Tunnel). By 2004, when the I-495 unsolicited project proposal was made, the State appointed an independent review panel, which ended up voting in favor of the private consortium proposal.

The enabling authority should also help to define the purpose of applying pricing. For example, there is an inherent difference in toll structure if the objective is to maximize revenue rather than to manage traffic flow. Gaining some clear intent from the legislature can assist in designing a congestion pricing project that will meet political expectations.

There is substantial variation in political structures among regions in North America. The differences are more disparate when comparing North America to Europe, where most of the zone-based congestion pricing projects have been implemented.

In general, there are few international reference points from which the United States can obtain relevant lessons learned. However, as shown in the text box above, the successes of zone-based programs in London and Stockholm offer some perspectives.

Most regions have looked to define agency and political roles to avoid infighting. The biggest political concern seems to be "Who is going to be able to control the key decisions?" Some of this concern stems from the trust issue described previously. Agencies that have made realistic promises about congestion pricing impacts have helped to develop additional trust with political bodies.

Obtaining Authority for Zone-based Congestion Pricing Schemes

Zone-based pricing schemes have evolved from a variety of legislative beginnings. In **London** there was strong national support and enabling legislation. In London the Greater London Authority Act 1999 established the authority of the mayor. This was followed by the 2000 Transport Act, which expanded pricing authority to the rest of the country except for Scotland. This legislation paved the way for implementing the London pricing program.

In **New York City**, the possibility of a UPA grant encouraged City Hall to propose congestion pricing in April 2007, and UPA deadlines accelerated its consideration. Once the Mayor announced his plan, work proceeded quickly with the City leading a multiagency staff group, and then through a Commission set up by the legislature. While the regional groups proceeded well with the planning effort, City Council approved the plan and the State Senate was likely to do so as well, but the plan failed to gain support in the State Assembly. The primary legislative concern was the perceived lack of benefits to drivers (e.g., reduced congestion) relative to the cost of the charge. This experience pointed out the difficulty of achieving statewide support for a specific regional transportation program.

In **Sweden**, congestion charging was considered in both Stockholm and Gothenburg. The **Stockholm** plan was first initiated in the early 2000s. Initially there was no legislative authority—but local agencies pressed the national legislature to get involved. The Stockholm pricing scheme was approved as a 'trial' for 7 months during a 3-year period. Social Democrats took away some of the political heat by holding a referendum after the 7-month trial. Stockholm still needed national legislation (termed a "national tax" under Swedish law) to fully authorize the pricing scheme. Local agencies needed to ask for this authority, so the city of Stockholm proceeded to make the request, even though the leader of the city was not supportive. The referendum passed in 2007, providing funds to pay to construct a new ring road plus a small amount allocated to transit. The city came on board after the trial period ended.

Gothenburg Initiated planning for a cordon pricing scheme in 2009. The city's goal was to help fund needed infrastructure, including for transit. They used a similar planning process as Stockholm but had new national legislation that delegated the congestion pricing decisions to a regional authority. The Swedish National Road Administration was supportive of the scheme along with the City of Gothenburg. This support overcame the initial resistance to the idea by the Swedish Department of Growth. The pricing scheme opened on January 1, 2013.

Parking Pricing Decisionmaking

Parking pricing decisionmaking tends to be more localized within a city or other municipal agency. Parking has always been a very visible, contentious and political issue within cities, given the localized relationship between parking, local businesses, and municipal funding. The advent of parking pay stations and potential for variable on-street meter parking rates creates a new institutional dynamic—cities now have more control over the hourly parking rates charged in different areas within a city. The complication is that cities control most on-street and some off-street parking, but most off-street parking is privately owned and operated. In the *SFpark* project, the San Francisco Metropolitan Transit Authority was able to obtain legislative approval early in the process through the city council rather than obtaining state approvals.

The FHWA primer *Contemporary Approaches to Parking Pricing* (2012) provides additional insights into the complexities of implementing parking pricing programs, including examples from **San Francisco**, **Seattle**, **New York City**, **Chicago**, **Los Angeles**, **Washington, DC**, **Ventura, CA** and **Aspen, CO**. The primer is available at *http://www.ops.fhwa.dot.gov/publications/fhwahop12026/index.htm*.

Organization

The implementation of congestion pricing projects often requires a different organizational structure than typically found in State DOTs and municipal governments. The introduction of a pricing mechanism means that revenues must be collected and distributed. Organizational issues include the degree of agency cooperation, structure and staffing, and the development of interagency agreements.

As previously stated, defining clear agency responsibilities facilitates a clean organizational structure. For priced managed lanes, agencies new to tolling must set up a tolling organization either within the State DOT or in the form of a separate toll agency. When private companies are involved, establishing clear responsibilities is essential to ensure that the public and private staff can work together as a team. For zone-based pricing, various State, regional, and/or municipal governments can be involved. Conversely, parking pricing is typically controlled by a municipal government, so the organization is more defined.

Organizations that can speak with one voice fare better in policy discussions with decisionmakers and the public. The complexities of congestion pricing projects can be smoothed out by having a clear message that is echoed by all of the parties involved. This can be achieved by educating the project team on the objectives of the project. If this is done early in the project planning phase, it will help to create a cohesive external project message and also keep the project team focused on the most important project needs.

A successful congestion pricing organization establishes an internal structure that involves the right people. This includes defining an internal group set

> **Key Points**
> - Define project roles – agency internal/external, and private sector
> - Get the right PEOPLE, not just the right AGENCY
> - Make sure everyone understands the project objectives
> - The agency that's going to operate the congestion pricing program should be the one that plans and designs it

apart from normal departments that is focused on the congestion pricing project. Since pricing projects are often multidisciplinary, agencies with internal departmental "silos" can face challenges associated with having the right staff involved at the right times. For example, accounting and finance staff are usually in different departments, and may not have normal lines of communication with transportation staff. Assigning the right staff from various departments to the congestion pricing team will help to break down the organizational silos. To facilitate cross-department coordination, some congestion pricing projects have developed specific line authorities among departments using internal agreements.

It is one thing to set up an organizational structure that looks good on paper. It is another to get the right people involved. Congestion pricing projects benefit from having staff who are committed to the project and who understand how the pricing of transportation facilities changes the dynamics of a transportation system. Part of this challenge is to form a team of professionals who recognize the need to become more "customer-focused" due to the introduction of toll transactions and additional questions generated by the public. Meeting this requirement may involve hiring or reassigning staff with customer skills into the congestion pricing team.

Many congestion pricing projects also involve close coordination between the implementing agency (e.g. State DOT) and transit providers. Buses may operate on the priced facility with special rules, which must be established up-front. In some cases, transit agencies might also be recipients of pricing revenues that will affect the delivery of bus services.

The private sector has become involved in congestion pricing projects through P3 agreements. Injecting private sector interests and personnel into a traditional public agency structure can lead to organizational challenges, as illustrated on the Virginia **495 Express Lanes** project example (see text box on next page). One challenge with P3 projects is that public agencies do not negotiate like private sector firms. Internal public agency structures and legislative provisos can limit the latitude of staff to make agreements that fully maximize the public benefits of a congestion pricing project. It is important to understand these limits when developing the P3 agreements and setting up the public/private organizations that will design and implement the project.

Organizational Strategies for Priced Roadways

Roadway pricing projects often pass through multiple jurisdictions and require close coordination among State, regional, and local agencies.

In **Miami**, the I-95 project involved two Florida Department of Transportation (FDOT) districts, the Florida Turnpike Authority, two Counties, and two Metropolitan Planning Organizations (MPOs). There was no formal "agreement" between FDOT and the Turnpike Authority, but staff were given specific authority during project planning, construction and operation. The agencies are working together well without official agreements.

In **Atlanta**, the primary agencies designated "lead" responsibilities split by topic: Design, Procurement, Project Management, Policy, and Finances.

In **Minnesota**, the Minnesota Department of Transportation (MNDOT) was the project lead, but the State also created a project management team consisting of a local consultant, a national consultant, and the Humphrey Institute. This management team became the "independent" face of the project and helped to dispel the perception that MNDOT was forcing tolling onto the public. In essence, this team held MNDOT at arm's length on the project so the agency did not appear to be primary champion. This proved to be a very effective approach.

In **Los Angeles**, the Los Angeles Metro Chief Executive Officer held a weekly staff meeting and included an agenda item at every other meeting to discuss how the senior staff were coordinating on the project. This organization set forth a culture that promoted full integration of the program team rather than just cooperation for a single project. In turn, the leadership style broke down internal organizational silos, which can hinder implementing pricing strategies. One challenge was to develop a cooperative agreement defining roles and responsibilities between LA Metro and Caltrans, but this agreement was achieved.

In **Seattle**, the SR 520 Bridge tolling project has been a joint effort between WSDOT, King County Metro (i.e. transit service provider), and the Puget Sound Regional Council. These agencies collaborated on the Urban Partnership proposal and have remained active partners during implementation of the pricing project. WSDOT is the overall lead for the project, which requires coordination between the Toll Division and the Northwest Region, which is constructing the new floating bridge and connecting freeway segments.

Integrating Public and Private Sector Organizations

The **Virginia 495 Express Lanes** created a new organizational structure within the Virginia Department of Transportation (VDOT). Both the public and private sector partners needed to be integrated into a team, so, to accomplish this, VDOT set up an internal structure to manage the I-495 project. A 2002 VDOT reorganization created two divisions: the Innovative Project Delivery Division, and the Innovative Finance Division. This structure worked reasonably well, but the two divisions were still somewhat independent. In December 2010, the Secretary created the Office of Transportation Public Private Partnerships (OTP3), which then oversaw the two innovative divisions.

Organizationally, VDOT hired a General Engineering Consultant, which was responsible for interacting with all of the affected parties. VDOT and the consulting team shared many responsibilities for design, but VDOT retained approval authority for monetary expenditures and task order changes. Given the newness of the P3 process, the various divisions within VDOT had different understandings of what was included in the contract. In hindsight, staff indicated that it would have been very helpful to make sure that all VDOT staff clearly understood the contract with the private concessionaire. This could have prevented some misunderstandings and avoided some change orders. Complicating the issue was the fact that the concessionaire initially did not know much about the Northern Virginia planning process. To resolve this, VDOT hired a seasoned professional from Fairfax County to manage the overall planning process.

Some VDOT divisions and personnel were less willing to give up design and implementation control to the concessionaire. There was a perception that the State would ultimately be responsible for the project, so relinquishing control was difficult. In the early months, moving beyond the normal processes the VDOT was used to was problematic, even though the senior department managers embraced the overall project. Some of the challenges were also organizational: at the start of the project, VDOT handled the P3 as part of its "normal operations," which meant that the few people involved made the project a priority. Subsequently, VDOT put one person in charge and set up a separate office for P3. More streamlined procedures and processes followed, the level of trust strengthened between the parties, and clear expectations were articulated on both sides.

Organizational Challenges with Zone-based Pricing

For zone-based pricing schemes, having regional political autonomy facilitates making decisions. **London** is the prime example of having a central government that controlled the planning, design, and implementation of the congestion pricing scheme.

New York City had similar aspirations, but the greater institutional complexity made it complicated to gain public and political support. While the city led a strong multiagency working group including city, State and transit agencies, the Metropolitan Transportation Authority (MTA) would have spent the money. Despite a "lockbox" provision in the proposed authorizing legislation, many questioned whether the revenues would be spent as intended, undercutting support for the plan.

Parking Pricing Organization

Parking pricing has many of the same institutional issues associated with multiple agency involvement, but parking also has some unique challenges. Institutional issues follow who controls the parking and who pays the parking. A city controls only a certain amount of the available parking; most of the parking in key congestion areas is privately owned and operated. This reality requires close coordination between the agency and the private sector. The use of in-pavement parking sensors and potential inclusion of private parking operators introduces complexities for negotiation of a parking pricing scheme, although this usually would not change the overall program organization.

Another complication is the competing needs for curb space on city streets. Within an urban area, the curb space accommodates parking and many other uses—commercial vehicle loading, taxis, bikes, food trucks, etc. As a result, parking is highly visible and political, leading to local officials making parking decisions based on qualitative factors that don't necessarily benefit the city's overall parking program.

SFpark in San Francisco is the first parking management project to set parking rates in a very transparent fashion using quantitative, real-time parking demand information. SFMTA controls on-street parking as well as several off-street public parking garages. This gives the program flexibility to price both the on- and off-street parking to achieve the project goals.

Before the *SFpark* project, it cost more to park in a garage than a meter, with a result that garages were underutilized. The *SFpark* shifted 14 of the 20 SFMTA public parking garages back to their original purpose – short term parking rather than commuter parking. *SFpark* lowered most hourly rates (at the vast majority of garages and times) but increased the cost of commute (early bird/daily/monthly) parking. This action affected some commuters who were used to all of the "early bird" parking specials, and seems to have influenced the market price or pricing structure at surrounding privately owned/operated garages.

One technique used to define the organization clearly is the agency agreement, which can either be intra- or interagency. This agreement can distinguish between the major funding and implementing organizations, and those partners who may have small but unique roles. In the case of priced managed lanes, several agreements may be needed throughout a corridor to deal with changing jurisdictional boundaries, or in the case of the Miami I-95 HOT lanes, changing transit providers.

Planning Process

In many ways, the planning process involved in implementing a congestion pricing project is similar to other major transportation projects: set the objectives early, include all affected groups, and assume the need for continual evaluation. Because of the newness of the congestion pricing concept, decisionmakers and the public appreciate having a clear set of objectives that the project is trying to achieve.

The various agencies and private partners may have competing objectives that should be resolved early in the planning process. For example, a State DOT may want a priced managed lane to generate sufficient revenue to cover operating costs along with some of the capital costs. The resulting performance standards could prioritize toll revenue over freeway operations. This objective may conflict with a transit agency's goal to ensure free-flowing travel (for buses) in the managed lane. Especially during the early HOV to HOT conversion projects, transit agencies operating on the HOV lanes tended to view the conversion to HOT lanes with different objectives than those of the implementing agency. Similarly, private funding partners typically want to make sure that they obtain a reasonable profit, an objective that may conflict with an implementing agency's goal to allow free travel for HOVs and transit, or to make sure that the lane(s) operate at a certain level of service. Successful priced roadway projects have involved all organizations throughout the process, including setting the project objectives/policies. Setting clear objectives is also important for zone-based and parking pricing programs, as illustrated in the text boxes on the following pages.

Once the objectives have been set, the challenge is to create a professional team based on essential disciplines to implement the planning process. For example, in

> **Key Points**
> - Set up clear objectives (e.g. revenue vs. operations)
> - Create clear planning process
> - Include all affected groups, including construction, operations, enforcement
> - Plan for continual evaluation- lessons learned
> - Don't forget the equity issues

Atlanta, the team was organized around the following functions: policy, finance, outreach, environmental, design, tolls, enforcement, and transit. Members working in each of these functions had specific roles within the planning process, but under the auspices of an overall team.

Several agencies cited the need for a systems approach to congestion pricing planning, even if the immediate plan is at corridor or subarea level. For example, some planning and design decisions (such as HOV eligibility definitions, toll rates, design treatments, and technology choice) might make sense for a single priced roadway corridor, but they may not make sense from a system perspective. In Northern Virginia, managed lanes evolved in different corridors with different HOV occupancy rules. As a result, the toll transponders include a 3-way switch to adapt to the different rules on I-495, I-395, and I-66. Some additional examples are provided below. Consistency and simplicity are good advice for planners to consider when planning for a congestion pricing program.

Once the initial planning is done and the project is close to implementation, the question remains "will the

Zone-Based Pricing Objectives

There are several examples of how the definition of objectives influenced the design and success of zone-based congestion pricing schemes:

London – The overall focus of the London scheme was to reduce delay for all travelers within central London. This goal led to these objectives:
- Create less delay for remaining traffic
- Reallocate road space to environment
- Improve crossing times for pedestrians
- Provide modal choices for people

Within this context, most people perceived that they would see some benefits from the pricing scheme.

New York City – The congestion pricing goals were consistent with the city's comprehensive sustainability plan, which emphasized congestion reduction, cleaner air, and increased funding for mass transit improvements. These goals led to these specific objectives:
- Reduce Vehicle Miles Traveled in downtown by 6.7%
- Reduce traffic delays

Raise money for transit. Since most of the toll funds would go towards transit rather than roadway improvements, some drivers felt that the indirect travel time savings would not offset the cost of the toll.

Stockholm – The goal of the original Stockholm cordon pricing scheme (from the 1990's) was focused on generating revenue to fund a transportation package of improvements. Attempts to subsequently market this program under the guise of environmental benefits and reduced congestion helped lead to its failure. Several years later, a new pricing scheme was proposed with the following objectives:
- Improve the environment (air quality, CO_2)
- Improve traffic flow entering the city center
- Raise transportation funds

This cohesive set of objectives provided a stronger basis for public support and its implementation.

Manchester (UK) - The Manchester plan did not specify clear objectives and benefits, which may have contributed to its lack of public and political support

---

project be successful?" Developing a thorough before/after planning study to actually measure the results of the congestion pricing project can help to answer many of the lingering questions raised by the public and decisionmakers. Some of these results can include the extent to which the project achieved the institutional goals set early in the planning process. Reporting lessons learned will transfer knowledge and allow others to apply these lessons to new project applications.

Project evaluation doesn't need to wait until after the project is implemented. In Los Angeles, the I-10/110 project team conducted an internal lessons-learned workshop right after the preliminary engineering phase to address some serious policy and design issues. This self-evaluation helped the project team to refocus on the key project objectives and enabled them to proceed into the next phase in a positive manner.

One of the most important planning issues relates to traveler equity; specifically, income equity.[3] The benefits of congestion pricing may not be distributed equally among all users, giving rise to the popular "Lexus Lane" perception. While research has shown that all income groups can receive benefits from congestion pricing projects and generally support the concept, the income equity issue is one that surfaces on almost every project. The planning process can help by identifying the im-

---

[3] Refer to Federal Highway Administration, *Income-Based Equity Impacts of Congestion Pricing, A Primer*, December 2008.

Planning for Roadway Pricing

Roadway pricing projects share a need for setting an understandable planning process with clear objectives. As an example, the **Atlanta** I-85 project team asked the following question at the start of and during the project development: "What are the project fundamentals?" "What is the purpose of this project? "What must be accomplished?" Answering these questions led to a definition of key objectives and the identification of which team members would be responsible for each action.

In **San Diego**, the I-15 project had a specific focus on lane management and maintaining Level of Service C. These clear objectives helped to frame inter-agency expectations.

The **Los Angeles** I-10/110 team laid out specific project objectives, but during the planning stage team members developed some competing expectations around the treatment of HOV 2+ vs. 3+ users in the priced lanes. An original objective included allowing HOV 3+ vehicles into the HOT lanes for free, while HOV 2 vehicles would pay the toll. This objective generated debate around the local public policy that promotes carpooling at all levels.

In **Seattle** the HOV eligibility issue also arose as a policy debate relative to proposed express toll lanes on I-405 and for tolling on the SR 520 bridge. On I-405, an existing HOV-2 lane would be converted to dual HOT lanes, with HOV 3+ vehicles paying no toll. This planning decision affected both revenue estimates and carpool promotion within the region. The SR 520 corridor already had an HOV 3+ rule on the managed lane approaching the bridge. In this case, the planning decision to charge tolls to all HOVs (vanpools and buses are free) was relatively uncontroversial.

Injecting Private-Sector Initiatives into the Planning Process

In **Northern Virginia**, a complex planning process was already underway for the Capital Beltway (I-495) expansion when the unsolicited HOT lane proposal arrived from the concessionaire. At that time, the I-495 alternatives being considered would result in substantial right of way takes and high capital cost. The concessionaire proposal promised a smaller project footprint with a much lower price tag. It also brought HOV/HOT lanes and transit benefits into the corridor. After substantial review, this design concept was added as a new alternative and the EIS process was restarted and completed. Having the EIS process already underway streamlined the approvals and was likely seen as a positive factor by the private applicant.

---

pacts and benefits to each user group and design the project to maximize the benefits. For example, using a congestion pricing program to provide improved transit service can be a key to public acceptance and addresses many of the equity issues that arise. Other related strategies that can address the income equity issue include protections for low-income travelers, such as toll credits, exemptions, or other forms of monetary compensation.

Defining the pricing geographic area is an important congestion pricing planning issue. This is a particularly sensitive issue for zone-based pricing programs, where setting the pricing area boundaries defines who pays and who doesn't (a person pays when they cross the boundary). In general, larger boundaries are preferred, since they affect a fewer number of local trips that stay within the boundary. Examples from London and New York are described in the text box on the next page.

The influence area is also important for roadway pricing projects, where planners should consider the diversion effects of the priced roadway on parallel freeway and arterial facilities. In Seattle, tolls were implemented in 2011 on the SR 520 bridge across Lake Washington. The toll resulted in a diversion of traffic to a parallel bridge (I-90) four miles away. In anticipation of these effects, the State is considering adding tolls to I-90 to balance the traffic demands and smooth the traffic congestion effects.

For parking pricing applications, creating a large enough parking pricing zone to avoid parking competition issues is a planning consideration. In the *SFPark* program, the zone includes most of downtown San Francisco, which comprises a large parking market.

Setting the Pricing Area Creates Equity Issues

Zone-based congestion pricing schemes deal with a broad range of equity issues. This is because an artificial boundary is established, creating different classes of travelers—those who pay and those who don't. Issues surrounding **vertical equity** (also called **income equity**) certainly exist, since all travelers must pay. Setting the boundary location and size can therefore affect equity.

In **London**, income equity was not seen as a huge issue since transit is strong and bus riders would benefit the most by the increased transit service. Rather, the primary inequalities were seen as being **spatial**, with primary effects falling on businesses inside/outside of the boundary and near the boundary.

In the **New York City** situation, the proposed pricing scheme boundaries encompassed the Manhattan business district (initially, from 86 Street to the Battery, and later from 60th Street to the Battery). However, the plan also credited tolls paid against the congestion fee, with the result that far more drivers crossing from New Jersey would partially or completely escape the congestion fee, as compared with drivers entering from Queens, Brooklyn or northern Manhattan/the Bronx. This perceived spatial inequality was a major point of contention about the plan.

Photo source: 495 ExpressLanes, Transurban

Public Involvement

An inclusive public involvement plan can build support for congestion pricing projects. Public involvement requires a careful blending of informing and listening. There is a need for outreach to both decisionmakers and travelers, both of whom are dealing with a new transportation approach. Up-front education is important to articulate the objectives and benefits of the project in a way that is meaningful to the audience. The public outreach plan should try to reach out to all affected travelers, some of whom may only be occasional users of the congestion pricing scheme.

Since money is involved, the public will need to understand what they are receiving in exchange for paying a toll or paying more for a parking space. Similarly, decisionmakers need the same information to be able to explain the project to their constituents. Many congestion pricing projects also need to reach out to businesses that would either be affected by the pricing scheme (e.g., access to parking) or whose employees would need to pay for traveling to and from work.

Successful congestion pricing projects keep moving ahead. Given the newness of many congestion pricing strategies, most projects have found that it is important to persevere with the project message, even if there is not initial support. Setbacks can occur along the way, but keeping the momentum means that the project acceptance doesn't need to go back to zero if there is a failure. Minneapolis initially tried various ways to implement the MNPASS program, some with more success than others. Even with some early failures, MNDOT found that public understanding and support grew to a new plateau from which new initiatives could be launched. Once a project has crossed the basic threshold of success, there is an ongoing need to maintain and enhance the project messaging to keep the momentum building to avoid moving back to square one.

> **Key Points**
> - Articulate the objectives and benefits
> - Persevere, even if you don't have initial support
> - An inclusive public involvement plan can build support

For decisionmakers who are unfamiliar with a particular congestion pricing strategy, agencies have used a variety of interactive tools—workshops, expert panels, peer exchanges, study tours—as a way to engage decisionmakers in the process. Knowing that a strategy has been successfully implemented elsewhere provides some assurance to a decisionmaker that it could also be achieved within their community or region.

Much like any strategy that involves payment of money, the initial perception of congestion pricing within a community may be negative. A natural political reaction can be to hold a public referendum to gauge the level of community support. As described below, the few regions (all in Europe) that have held public referendums on a congestion pricing scheme prior to implementation have given mixed reviews to that strategy. Typically the public doesn't understand enough about the congestion pricing objectives and potential benefits to weigh the tradeoffs versus the perceived individual costs.

The outreach strategies are ultimately aimed at creating a knowledgeable traveling public and gaining support. In addition to providing timely and ongoing information to the public, some projects have taken proactive steps to put mitigations in place before pricing starts. In San Diego, SANDAG committed to enhanced bus service in the I-15 corridor to start shortly prior to the

> **Roadway Pricing Outreach Tips**
>
> Roadway pricing evokes strong initial public reactions, typically negative. Each of the operating HOT lane projects included extensive public outreach focused on telling the public what HOT lanes are, how they operate, and what would be the potential impacts or benefits.
>
> Some of the tips from these outreach efforts include:
>
> - Use messaging that emphasizes potential benefits to users, such as using the term "value pricing" and giving people a "choice." San Diego emphasized that "these are managed lanes that are not static and can be adapted over time."
> - Emphasize a regional strategy that diffuses the question "Why us? Why can't you do it somewhere else?" In Los Angeles, they branded the effort as a "program," not a "project."
> - Distinguish between "customer" and "driver." Roadway pricing creates new "customers" for the agency, while the term "driver" implies an impersonal and generic perspective.
> - Keep the outreach going after the project is open. In Minneapolis, there was initial skepticism of the I-394 HOT lanes by some communities that didn't see the potential benefits coming to them. Legislators and community leaders persuaded the public to give the project a chance. Once the I-394 project was in operation for a few months, people quickly saw the direct benefits to them as travelers and didn't witness any substantial negative effects. The outreach campaign used this information and positive opinion polls to further build public support.
> - Deal directly with community concerns. In Los Angeles, most of the negative reactions to the proposed I-10/110 HOT lanes were from legislative and elected officials' pushback. Anti-pricing elected officials had a negative perception of how the HOT lanes would affect equity within their communities. In reality, most of the low-income communities were less concerned about the HOT lanes and more concerned about using HOT lane revenues to improve transit service along the HOT lanes.

beginning of the HOT lane project. Similarly, as part of the SR 520 toll bridge project in Seattle, WSDOT worked with the regional transit agency to initiate a substantial increase in express bus service within the SR 520 corridor for several months leading up to the start of bridge tolling. This created an expanded transit market and helped to reduce traffic diversion effects. In addition, WSDOT provided traffic mitigation funds to local agencies to assist in dealing with the expected traffic diversion.

What about Holding a Public Referendum?

In Europe, leaders in charge of planning proposed zone-based pricing schemes grappled with the question of whether to ask the public its opinion of congestion pricing in the form of a referendum. Here are some observations:

- London – Mayor Livingstone said no vote was needed and moved ahead with the project. This points to the centralized decision-making for the London scheme.
- Edinburgh – The Scottish government gave approval for the pricing scheme contingent on gaining public support. It wasn't clear that the legislature required the city to go to referendum, but they did. The Pricing scheme lost by an 80/20 vote.
- Manchester – The city held a referendum, including residents of 10 local authorities. The process became very political. For example, local groups lobbied to support candidates if they committed to support the referendum. The vote also lost 80/20.
- Stockholm – The city implemented the pricing scheme first, and then held a referendum, which passed 53/47. Speculation was that implementing a "pilot" pricing scheme for several months helped to build public support.

Public referendums are always difficult to predict, especially dealing with an entirely new transportation and revenue concept that includes "tolls."

SFpark Public Outreach

SFpark conducted a 3-year outreach effort to educate the public, the private parking operators, and local businesses. The program has not been very controversial, since it was portrayed as a win-win solution, including making parking easier to find and promising fewer parking tickets.

The city emphasized that the program would focus on demand management rather than revenue generation. The "revenue neutral" objective was an important marketing message. The project team specifically avoided the use of obscure jargon such as "congestion pricing" or "traffic management."

Managing Costs and Revenues

> **Key Points**
> - Resolve local/regional interplay and how funds would be used
> - Keep the revenues close to home
> - Giving some revenues to transit helps with equity argument

Congestion pricing projects by their nature include revenues, either through tolls, fees, or parking charges. Proper accounting for these revenues, and how they will be spent, can play a major role in gaining public and agency support for the project. The interagency agreements previously discussed can be explicit regarding who will be paying, who receives the funds, and how the funds will be expended.

The other side of revenue is cost. Congestion pricing projects include a combination of capital costs and operating costs. Capital costs include the one-time costs for construction and equipment while operating costs cover ongoing staffing, administration, and maintenance expenses. Most congestion pricing project revenues are sufficient to cover most or all operating costs, but capital cost recovery is highly variable. This is especially true for HOV to HOT lane conversion projects, where the revenue stream is relatively low compared to the up-front capital cost to create the HOT lanes. As more multilane HOT facilities are being developed, there is potential to generate more toll revenue and the ability to offset more of the capital costs.

Funding is a major issue with public private partnership arrangements due to the intertwined nature of cost sharing vs. risk vs. revenues. Since many concessionaire contracts extend for 40 or more years, there is considerable risk involved for whichever party assumes the responsibility for covering project costs in the later years. If both parties share the same project objectives, there is a greater likelihood that costs and revenues will be equitably allocated.

Deciding how to allocate revenues often specifies who has control over the congestion pricing project. Situations that have arisen include the following:

- **Addressing potential revenue shortfalls during project ramp up.** The Seattle SR 167 Express Lane and Atlanta I-85 projects identified the need to prorate revenue expectations to account for the likely slow build-up of priced lane demand. Both of these projects were the first HOT lanes implemented within their respective regions.

- **Decide how to handle cross-state toll payments.** The Virginia I-495 Express Lanes are used by many travelers who reside in Maryland. One of the challenges the project has faced is how to obtain and allocate toll payments from Maryland drivers. Other priced lane projects also need to deal with out-of-state driver tolls and revenues.

- **Revenues are often legislatively mandated to be spent on tolled facility.** Many states include legislative provisions that mandate that toll revenues be spent on the tolled facility or other specified roadways. As more priced roadways are implemented within a region, these mandates may limit some flexibility in spending the revenues that are generated.

Revenues in most congestion pricing projects have been used largely to offset the ongoing Operations and Maintenance costs. Where there are excess revenues, there is a desire to both keep the revenues close to home as well as distribute funds to benefit a large enough constituency. As previously discussed, spending some funds on improved transit creates agency and public support. This has been the case on the I-15 managed lanes in San Diego and in the London cordon pricing scheme. A portion of parking revenues in San Francisco are also spent on downtown transit enhancements.

Examples of How Priced lane Projects Address the Revenue Issue

- **Atlanta** – During the I-85 project development, the planning team discussed future revenue allocation, but because the project was unlikely to have much excess revenue to deal with in the early years, the issue was successfully moved to the back burner. Each of the agencies did have a position on where the funds should go, so the topic will arise later as the project matures.

- **Miami** – The I-95 project team estimated that it would be 2031 before the HOT lanes are making a profit. The first priority for surplus revenue would go to operate the corridor. Subsequently, the legislature passed a law specifying that once the non-toll funding has been repaid, any remaining toll revenue would be used for the construction, maintenance, or improvement of any road on the State highway system within the county or counties in which the revenue-producing project is located.

- **San Diego** – The I-15 HOT lanes generate revenues in excess of operating expenses. SANDAG is required to cover its share of the Caltrans cost for operations of the express lanes based on the availability of revenue and the percentage of FasTrak traffic. Any excess revenue must be spent within the corridor and, although there is no specific revenue allocation requirement, historically, around 25 percent of net toll revenue has been spent on corridor transit investments. In 2012 this amounted to around $1 million annually. The funds are paid to transit from previous year revenues.

- **Minnesota** – The enabling legislation allowed the State to borrow from other capital program funds to implement a toll facility, but the project revenues must then be used pay back the fund from which the money was borrowed. In addition, the Law established that 50 percent of excess revenue (after capital recovery and operations) must to be used for transit enhancement and 50 percent must to be used for other corridor improvements. This provision added an incentive to support the HOV to HOT conversion.

- **Virginia** – With the private concessionaire involved in the I-495 project, negotiations about costs and revenue sharing became a focal point of the project agreement. After negotiations, VDOT agreed to share the costs as follows: VDOT would pay 30 percent, the concessionaire would pay 28 percent, and the remaining 42 percent would be borrowed with the private sector holding the risk. Over the years, if the project profits exceed a specified level, the agreement provides for a profit sharing split proportionally among the public and private funding partners.

Administrative Cost Issues with Zone-based Pricing Schemes

A big issue for zone-based congestion pricing schemes is management of administrative costs. Given the ongoing need to monitor traffic and process the revenues, administrative costs are typically in the range of 15-25 percent. The **City of London** wanted to make sure that its cordon pricing scheme worked well and focused extra attention on administration. As a result, the city was spending 40 percent of the pricing revenues on overhead costs. This percentage has been reduced as the program has matured and the cordon boundary was expanded.

Managing Parking Revenues

For parking pricing programs, a successful strategy has been to invest the parking revenue into neighborhoods where it's collected; this has been an objective in Washington, DC. Making this commitment builds agency credibility and gives the opportunity to show a direct link between increased prices and the service that's provided.

Parking revenue allocations in San Francisco are set by city charter to go to the SFMTA for funding of transit and related programs. Parking management rather than revenue generation was the primary objective of the *SFpark* program. As a result, allocation of revenues was not a big issue during the development of the program. The *SFpark* team did not promise revenue neutrality, but emphasized that some revenues would increase (meters) and others would decrease (citations/garages). During the test phase, they also publicized that the program would change parking prices as demand requires, thus affecting the overall revenue mix.

Implementation

Once the congestion pricing project has been planned and designed, it is time to start the project. Implementation addresses a number of topics, including construction and roll-out, day-to-day operations, contracting, and use of technology.

There is a natural tendency to roll out a project as soon as practical. This desire should be tempered by the need to ensure that the project will be as successful as possible. There aren't too many opportunities to try again, so it is important to "get it right" when the project opens up.

"Getting it right" can lead agencies to gravitate towards implementing straightforward, non-controversial projects as their first foray into congestion pricing. By making one project a success, then other opportunities may arise.

Several congestion pricing projects have been marketed as pilot projects, sometimes with a sunset clause tied to the results of a detailed evaluation of its effectiveness. The SR 167 HOT Lanes in Seattle was marketed as a pilot test, with relatively low expectations given regarding revenue generation. That particular corridor had relatively low HOV usage and moderate freeway congestion, so its selection as an HOV-to-HOT conversion project was relatively low risk. Its performance success and general lack of controversy has helped to build support for HOT lanes in other regional corridors. In Atlanta, the I-85 Express Lanes were billed as a "first phase," not a "demonstration project." This approach was successful in giving the impression that the express lanes were something new but part of a broader regional strategy.

There are many decisions to make prior to opening a congestion pricing program. For priced roadways, one of the most important decisions is to create a pre-opening policy for exempt vehicles (e.g. bus, HOV, emergency vehicles). These decisions will have been made months before, during the project design, but should be carefully communicated to operating personnel. Enforcement responsibilities also need to be clearly articulated to the enforcement agencies to which the public will be looking to ensure the integrity of the priced facility.

Ahead of opening, any new technology should be thoroughly tested. In today's fast-paced technological world, there is a strong tendency to use the newest field equipment and computer systems to provide project efficiencies. In reality, technology may outstrip practicality (and public acceptability), so the provider must balance technical innovation versus risk taking. Other areas for full testing are the user interfaces such as toll accounting procedures. Establishing convenient user payment options is one step that can help build community trust in the project.

Implementing parties should coordinate the timing and magnitude of any mitigation (e.g. actions to minimize impacts of diversion), new facilities (e.g. express lane ramps, park-and-ride lots) and complementary services (e.g. transit, ridesharing). Ideally, these actions

---

Key Points

- Set up the implementation process early in the project to ensure the program starts successfully
- Consider pilot projects
- Balance technical innovation vs. risk taking
- Make prudent use of technology

should take place concurrent with or before opening the congestion pricing project, to minimize the potential for negative reaction from travelers.

Once the project is up and running, the project team should closely monitor the daily operations. Flexibility is key in terms of being able to modify traffic management strategies or other actions in response to changing daily conditions.

Implementing a Parking Pricing Program

Parking pricing programs typically require an investment in parking sensor technology and data management. Often these technology changes create new relationships between a city and parking vendors. While the technology is important, there are many non-technical, often very simple, implementation issues that will need to be addressed, such as making sure that the parking spots are numbered so that a person knows which spot to pay for, perhaps with his/her cellphone.

The San Francisco *SFpark* program is taking advantage of new parking sensor technology and the city's expanded data management system to provide real-time information to travelers about available parking spaces and their prices. Regular interaction takes place between the SFMTA (lead), its meter shop and information technology departments, the private company that collects the meter funds, and various meter vendors. This is a case where the city is pushing the envelope on available technology but not getting too far ahead of the curve.

The *SFpark* program also allows the city to manage employee parking. Previously city vehicles were able to park for free at public parking facilities. Now, all city employees, except police, must pay. This change in policy, not directly tied to *SFpark*, reinforces the objective to have the pricing apply to everyone and is important for credibility, since the program is asking people to look at parking in a different way

The parking sensors installed by *SFpark* have created a new level of maintenance requirements for the city, which hopes to move away from in-pavement technology within a few years. New York City is also testing sensor technology and also exploring complementary technology, such as cameras and license plate readers for parking availability.

Roadway Pricing Implementation Challenges

Each roadway pricing project faces the basic challenges of starting the project successfully and ensuring that the priced facility works smoothly over time. However, there are some unforeseen issues that can arise.

For example, the Miami I-95 project faced a couple of particular implementation challenges

1. **Develop Pre-opening Policy for HOVs and Exempt Vehicles** – The I-95 project elected to develop and implement a carpool registration process, which was not in place prior to the HOT lane conversion. They also needed to establish a hierarchy for when certain vehicles are "bumped out" of the HOT lanes based on traffic conditions (for example: buses, 3+ HOV, Hybrids, 2+ HOV)

2. **Develop agreements between transit agencies (across county lines)** – Given the long length of the I-95 project, there are multiple transit agencies involved. Issues needing to be addressed during implementation included:
   a. What if a bus in one county breaks down in the other county?
   b. Can we unload from one bus to another bus (even though it's a different county's bus).
   c. How can we gain cooperation when one agency doesn't want to work with the other agency's bus?

The project team was able to have the transit agencies agree to help with transferring passengers and setting up rules regarding fares.

The **San Diego I-15** project went through various implementation stages. It started as a 2 year pilot project using a simple decal program. The agencies used this pilot to gauge success and then received broader legislative approval. Later on there were some concerns about the timing of the tolling to match the opening day of the express lane project. There were also many issues surrounding the customer accounts, such as payment of a monthly administrative fee. These were worked out over time. When the express lanes were extended from 8 to 20 miles, Caltrans needed to provide more access (initially just one access point, now 22). This major operational change led to higher operating and maintenance costs, including enforcement. Finally, with the I-15 express lanes extension having multiple access points, dynamic per-mile tolls were introduced to replace a single distance-based toll used on the original HOT lane project. These changes required more sophisticated design features but also provided the ability to process a multitude of tolls collected electronically at different points along the corridor.

The **Minnesota I-394** project provides a good example of needing to set up and operate an efficient and credible toll collection system. MNDOT hired an internationally reputable toll operator, which was provided space at a State-owned customer service center. An initial public concern was that the toll authority would price the facility to maximize revenue collection and then use the revenue how they pleased. Apart from State legislation that specifies how the funds should be collected and accounted for, MNDOT established strict protocols. The operator collects the money for MNDOT and deposits the money directly into a State account, eliminating the possibility of fraud. The operator was hired under a professional technical services agreement, so there are no incentives regarding the amount of revenue collected.

Lessons Learned

Congestion pricing injects new challenges into implementing transportation projects. These challenges are found both internally within an agency and externally in the public and among outside partners. By its definition, congestion pricing uses monetary pricing to help manage the transportation system. A byproduct of this pricing is revenue, which attracts a number of constituencies who have good ideas for how they could spend the money. Many of the congestion pricing challenges fall into a broad category of "institutional issues," which have been explored in this primer.

The following institutional issues tend to cut across the different types of congestion pricing projects: variably priced lanes, zone-based pricing, and parking pricing:

- **Strong leadership** – The newness of congestion pricing and typical skepticism by the public puts the onus on a project champion to guide the project though planning, design and implementation. Leaders can emerge from the political, civic, or private-sector.

- **Clear authority** – Most congestion pricing projects need some form of enabling legislation that should clearly identify who is in charge and what outcomes are expected. Clear authority is also needed within the project team.

- **Many Partners** – Pricing brings many new players to the transportation scene, including private sector investors. Traditional agencies and these new partners must be melded into a cohesive team. The organization should be structured to fit the needs of the project, not vice versa.

- **Know the objectives** – Agreeing to specific project goals and objectives up front in the process keeps everyone focused and creates a consistent message for the public and decisionmakers.

- **Educating the Public** – The public knows they will need to pay for something new, but what will they get in return? Educating the public on the purpose of the pricing scheme and what benefits they can receive is crucial to gaining support.

- **Know where the Money is Going** – Keeping the revenues "close to home" usually provides the most benefits to the people who are paying.

- **Get it Right** – Money is involved and the public is unsure, so work out the technology, accounting, and design issues before turning on the switch.

- **Flexibility** – There will likely be some crises with every project, but staying flexible helps to avoid surprises and allows for a more effective response.

These institutional issues point to a substantial shift in how traditional transportation agencies implement their projects. Suddenly, agencies now have many new "customers" who are using this new service. The move towards customer service requires a new organizational approach and a new type of leadership. As various types of congestion pricing projects continue to be deployed, organizations can build on the institutional ideas summarized in this primer to make their programs more successful.

References and Resources

REFERENCES

1. Federal Highway Administration. (2012, May). Contemporary Approaches to Parking Pricing: A Primer (FHWA Pub. No. FHWA-HOP-12-026). Washington, DC.
2. Gudmundsson, Henrik et al. (2009, March). Framing the role of Decision Support in the case of Stockholm Congestion Charging Trial. Elsevier Transportation Research Vol 43A, Issue 3.
3. Peters, Jonathan and Cameron Gordon. (2009, April). Results Not Guaranteed: A Tale of Road Pricing in New York and London. Journal of Urban Technology, 16:1,113-131.
4. Schaller, Bruce. (2010, March) New York City's Congestion Pricing Experience and Implications for Road Pricing Acceptance in the United States. Elsevier Transportation Research Vol 17.
5. Gordon, Cameron and Rich Flanagan. (2012, January). The Politics of Urban Congestion Pricing: Cautionary Tales from New York. Transportation Research Board Annual Meeting Compendium.

RESOURCES

The following individuals were interviewed to discuss institutional issues related to specific congestion pricing programs.

Project	Organization	Persons Interviewed
Variably Priced Lanes		
I-394, I-35W (Minneapolis, MN)	Minnesota Department of Transportation	Ken Buckeye
I-95 (Miami, FL)	Florida Department of Transportation	Rory Santana Jeff Weidner
I-495 Capital Beltway (Northern Virginia)	Virginia Department of Transportation Transurban, Inc.	Malcolm Kerley Ken Daley
I-15 (San Diego, CA)	San Diego Association of Governments	David Schumacher Sam Johnson
I-10, I-110 (Los Angeles, CA)	Los Angeles County Metropolitan Transportation Authority	Stephanie Wiggins
I-85 (Atlanta, GA)	Georgia State Road and Tollway Authority	Patrick Vu

Project	Organization	Persons Interviewed
Variable Tolls on Entire Roadways		
SR 520 (Seattle, WA)	Washington State Department of Transportation	Jennifer Charlebois
Zone-based Charges		
London and United Kingdom	Leeds University	Anthony May
London and New York City	University of Canberra	Cameron Gordon
New York City	New York City Department of Transportation	Tom Maguire
Stockholm and Gothenburg, Sweden	WSP Group, Sweden	Dirk van Amelsfort, Karin Brundell-Freij
Parking Pricing		
SF Park (San Francisco, CA)	San Francisco Municipal Transportation Authority	Jay Primus
The FHWA Primer *Contemporary Approaches to Parking Pricing* (2012) provides additional insights into the complexities of implementing parking pricing programs, including examples from **Seattle, New York City, Chicago, Los Angeles, Washington, DC, Ventura, CA** and **Aspen, CO**.		

The FHWA conducted a peer exchange in Washington, DC (May 2012) to discuss the range of institutional issues associated with a range of congestion pricing projects. The following professionals participated in the peer exchange.

Federal Highway Administration
- Angela Jacobs
- Wayne Berman
- Alan Greenburg

Georgia State Road and Toll Authority
- Patrick Vu

Florida DOT
- Rory Santana

Los Angeles County MTA
- Stephanie Wiggins

Bay Area MTC
- Lisa Klein

New York City DOT
- Bruce Schaller

San Francisco MTA
- Jay Primus

Washington State DOT
- Rob Fellows

Sound Transit (Seattle)
- Jim Edwards

San Diego Association of Governments (SANDAG)
- Dave Schumacher

Minnesota DOT
- Nick Thompson

Transurban
- Ken Daley

D'Artagnan Consulting
- Jack Opiola

SAIC
- Myron Swisher

Fehr & Peers
- Don Samdahl

www.ingramcontent.com/pod-product-compliance
Lightning Source LLC
Chambersburg PA
CBHW081805170526
45167CB00008B/3328